July 17, 2012

CYBERSECURITY

Challenges in Securing the Electricity Grid

Highlights of GAO-12-926T, a testimony before the Committee on Energy and Natural Resources, U.S. Senate

Why GAO Did This Study

The electric power industry is increasingly incorporating information technology (IT) systems and networks into its existing infrastructure (e.g., electricity networks, including power lines and customer meters). This use of IT can provide many benefits, such as greater efficiency and lower costs to consumers. However, this increased reliance on IT systems and networks also exposes the grid to cybersecurity vulnerabilities, which can be exploited by attackers. Moreover, GAO has identified protecting systems supporting our nation's critical infrastructure (which includes the electricity grid) as a governmentwide high-risk area.

GAO was asked to testify on the status of actions to protect the electricity grid from cyber attacks. Accordingly, this statement discusses (1) cyber threats facing cyber-reliant critical infrastructures, which include the electricity grid, and (2) actions taken and challenges remaining to secure the grid against cyber attacks. In preparing this statement, GAO relied on previously published work in this area and reviewed reports from other federal agencies, media reports, and other publicly available sources.

What GAO Recommends

In a prior report, GAO has made recommendations related to electricity grid modernization efforts, including developing an approach to monitor compliance with voluntary standards. These recommendations have not yet been implemented.

View GAO-12-926T. For more information, contact Gregory C. Wilshusen at (202) 512-6244 or wilshuseng@gao.gov or David C. Trimble at (202) 512-3841or trimbled@gao.gov.

What GAO Found

The threats to systems supporting critical infrastructures are evolving and growing. In testimony, the Director of National Intelligence noted a dramatic increase in cyber activity targeting U.S. computers and systems, including a more than tripling of the volume of malicious software. Varying types of threats from numerous sources can adversely affect computers, software, networks, organizations, entire industries, and the Internet itself. These include both unintentional and intentional threats, and may come in the form of targeted or untargeted attacks from criminal groups, hackers, disgruntled employees, nations, or terrorists. The interconnectivity between information systems, the Internet, and other infrastructures can amplify the impact of these threats, potentially affecting the operations of critical infrastructures, the security of sensitive information, and the flow of commerce. Moreover, the electricity grid's reliance on IT systems and networks exposes it to potential and known cybersecurity vulnerabilities, which could be exploited by attackers. The potential impact of such attacks has been illustrated by a number of recently reported incidents and can include fraudulent activities, damage to electricity control systems, power outages, and failures in safety equipment.

To address such concerns, multiple entities have taken steps to help secure the electricity grid, including the North American Electric Reliability Corporation, the National Institute of Standards and Technology (NIST), the Federal Energy Regulatory Commission, and the Departments of Homeland Security and Energy. These include, in particular, establishing mandatory and voluntary cybersecurity standards and guidance for use by entities in the electricity industry. For example, the North American Electric Reliability Corporation and the Federal Energy Regulatory Commission, which have responsibility for regulation and oversight of part of the industry, have developed and approved mandatory cybersecurity standards and additional guidance. In addition, NIST has identified cybersecurity standards that support smart grid interoperability and has issued a cybersecurity guideline. The Departments of Homeland Security and Energy have also played roles in disseminating guidance on security practices and providing other assistance.

As GAO previously reported, there were a number of ongoing challenges to securing electricity systems and networks. These include:

- A lack of a coordinated approach to monitor industry compliance with voluntary standards.
- Aspects of the current regulatory environment made it difficult to ensure the cybersecurity of smart grid systems.
- A focus by utilities on regulatory compliance instead of comprehensive security.
- A lack of security features consistently built into smart grid systems.
- The electricity industry did not have an effective mechanism for sharing information on cybersecurity and other issues.
- The electricity industry did not have metrics for evaluating cybersecurity.

United States Government Accountability Office

Chairman Bingaman, Ranking Member Murkowski, and Members of the Committee:

Thank you for the opportunity to testify at today's hearing on the status of actions to protect the electricity grid from cyber attacks.

As you know, the electric power industry is increasingly incorporating information technology (IT) systems and networks into its existing infrastructure (e.g., electricity networks including power lines and customer meters). This use of IT can provide many benefits, such as greater efficiency and lower costs to consumers. Along with these anticipated benefits, however, cybersecurity and industry experts have expressed concern that, if not implemented securely, modernized electricity grid systems will be vulnerable to attacks that could result in widespread loss of electrical services essential to maintaining our national economy and security.

In addition, since 2003 we have identified protecting systems supporting our nation's critical infrastructure (which includes the electricity grid) as a governmentwide high-risk area, and we continue to do so in the most recent update to our high-risk list.[1]

In my testimony today, I will describe (1) cyber threats facing cyber-reliant critical infrastructures,[2] which include the electricity grid, and (2) actions taken and challenges remaining to secure the grid against cyber attacks. In preparing this statement in July 2012, we relied on our previous work in this area, including studies examining efforts to secure the electricity grid and associated challenges and cybersecurity guidance.[3] (Please see the related GAO products in appendix I.) The products upon which this

---

[1]GAO's biennial high-risk list identifies government programs that have greater vulnerability to fraud, waste, abuse, and mismanagement or need transformation to address economy, efficiency, or effectiveness challenges. We have designated federal information security as a governmentwide high-risk area since 1997; in 2003, we expanded this high-risk area to include protecting systems supporting our nation's critical infrastructure—referred to as cyber-critical infrastructure protection, or cyber CIP. See, most recently, GAO, *High-Risk Series: An Update*, GAO-11-278 (Washington, D.C.: February 2011).

[2]Federal policy established 18 critical infrastructure sectors. These include, for example, banking and finance, communications, public health, and energy. The energy sector includes subsectors for oil and gas and for electricity.

[3]GAO, *Critical Infrastructure Protection: Cybersecurity Guidance Is Available, but More Can Be Done to Promote Its Use*, GAO-12-92 (Washington, D.C.: Dec. 9, 2011), and *Electricity Grid Modernization: Progress Being Made on Cybersecurity Guidelines, but Key Challenges Remain to be Addressed*, GAO-11-117 (Washington, D.C.: Jan. 12, 2011).

statement is based contain detailed overviews of the scope of our reviews and the methodology we used. We also reviewed documents from the Federal Energy Regulatory Commission, the North American Electric Reliability Corporation, the Department of Energy, including its Office of the Inspector General, and the Department of Homeland Security Industrial Control Systems Cyber Emergency Response Team, as well as publicly available reports on cyber incidents. The work on which this statement is based was performed in accordance with generally accepted government auditing standards. Those standards require that we plan and perform audits to obtain sufficient, appropriate evidence to provide a reasonable basis for our findings and conclusions. We believe that the evidence obtained provided a reasonable basis for our findings and conclusions based on our audit objectives.

Background

The electricity industry, as shown in figure 1, is composed of four distinct functions: generation, transmission, distribution, and system operations. Once electricity is generated—whether by burning fossil fuels; through nuclear fission; or by harnessing wind, solar, geothermal, or hydro energy—it is generally sent through high-voltage, high-capacity transmission lines to local electricity distributors. Once there, electricity is transformed into a lower voltage and sent through local distribution lines for consumption by industrial plants, businesses, and residential consumers. Because electric energy is generated and consumed almost instantaneously, the operation of an electric power system requires that a system operator constantly balance the generation and consumption of power.

Figure 1: Functions of the Electricity Industry

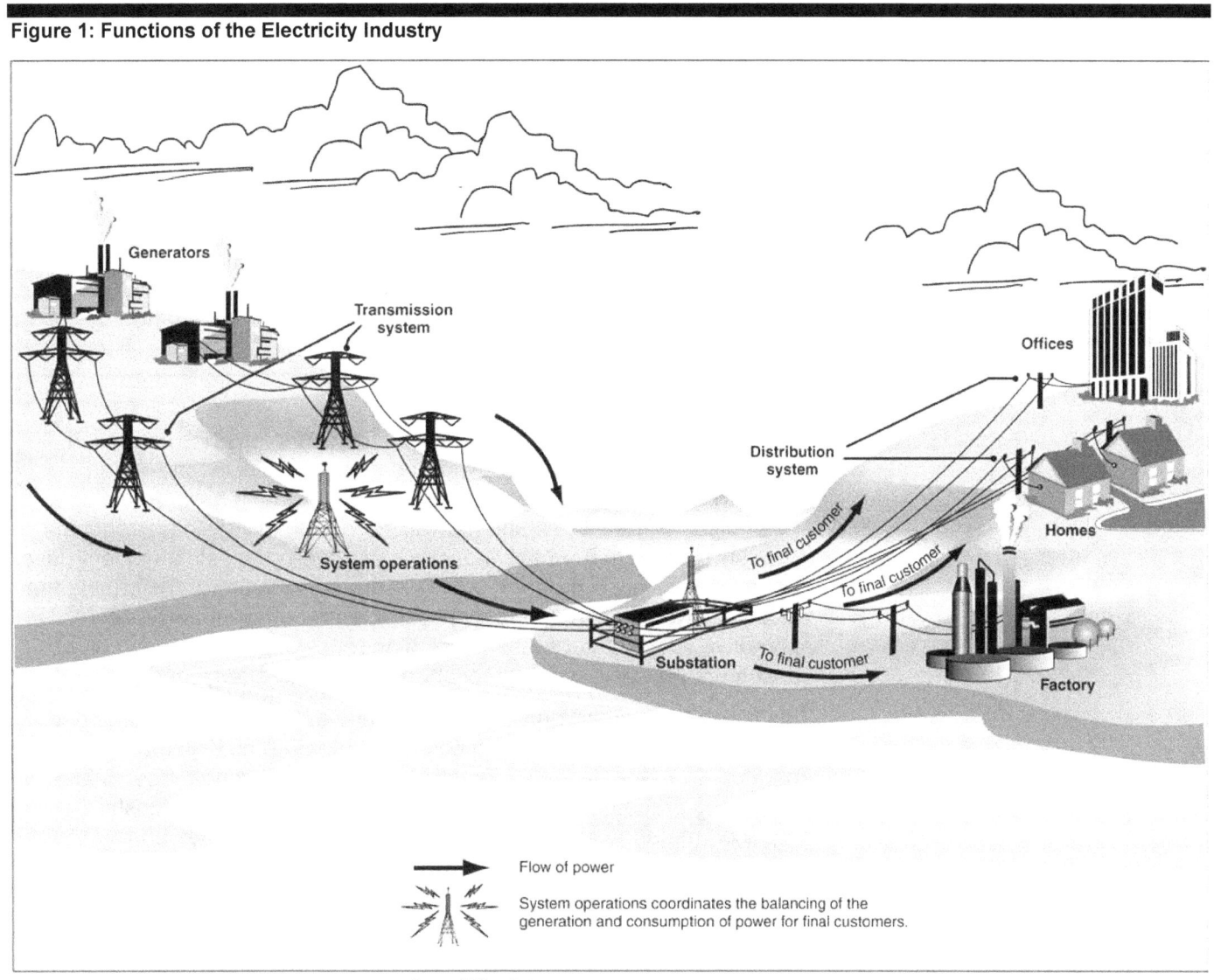

Source: GAO analysis.

Utilities own and operate electricity assets, which may include generation plants, transmission lines, distribution lines, and substations—structures often seen in residential and commercial areas that contain technical equipment such as switches and transformers to ensure smooth, safe flow of current and regulate voltage. Utilities may be owned by investors, municipalities, and individuals (as in cooperative utilities). System operators—sometimes affiliated with a particular utility or sometimes independent and responsible for multiple utility areas—manage the electricity flows. These system operators manage and control the generation, transmission, and distribution of electric power using control systems—IT- and network-based systems that monitor and control sensitive processes and physical functions, including opening and closing circuit breakers.[4] As we have previously reported, the effective functioning of the electricity industry is highly dependent on these control systems.[5] However, for many years, aspects of the electricity network lacked (1) adequate technologies—such as sensors—to allow system operators to monitor how much electricity was flowing on distribution lines, (2) communications networks to further integrate parts of the electricity grid with control centers, and (3) computerized control devices to automate system management and recovery.

Modernization of the Electricity Infrastructure

As the electricity industry has matured and technology has advanced, utilities have begun taking steps to update the electricity grid—the transmission and distribution systems—by integrating new technologies and additional IT systems and networks. Though utilities have regularly taken such steps in the past, industry and government stakeholders have begun to articulate a broader, more integrated vision for transforming the electricity grid into one that is more reliable and efficient; facilitates alternative forms of generation, including renewable energy; and gives consumers real-time information about fluctuating energy costs.

This vision—the smart grid—would increase the use of IT systems and networks and two-way communication to automate actions that system operators formerly had to make manually. Electricity grid modernization is an ongoing process, and initiatives have commonly involved installing advanced metering infrastructure (smart meters) on homes and commercial buildings that enable two-way communication between the utility and customer. Other initiatives include adding "smart" components to provide the system operator with more detailed data on the conditions of the transmission and distribution systems and better tools to observe the overall condition of the grid (referred to as "wide-area situational awareness"). These include advanced, smart switches on the distribution system that communicate with each other to reroute electricity around a

---

[4]Circuit breakers are devices used to open or close electric circuits. If a transmission or distribution line is in trouble, a circuit breaker can disconnect it from the rest of the system.

[5]GAO, *Critical Infrastructure Protection: Multiple Efforts to Secure Control Systems Are Under Way, but Challenges Remain*, GAO-07-1036 (Washington, D.C.: Sept. 10, 2007).

troubled line and high-resolution, time-synchronized monitors—called phasor measurement units—on the transmission system.

The use of smart grid systems may have a number of benefits, including improved reliability from fewer and shorter outages, downward pressure on electricity rates resulting from the ability to shift peak demand, an improved ability to shift to alternative sources of energy, and an improved ability to detect and respond to potential attacks on the grid.

Regulation of the Electricity Industry

Both the federal government and state governments have authority for overseeing the electricity industry. For example, the Federal Energy Regulatory Commission (FERC) regulates rates for wholesale electricity sales and transmission of electricity in interstate commerce. This includes approving whether to allow utilities to recover the costs of investments they make to the transmission system, such as smart grid investments. Meanwhile, local distribution and retail sales of electricity are generally subject to regulation by state public utility commissions.

State and federal authorities also play key roles in overseeing the reliability of the electric grid. State regulators generally have authority to oversee the reliability of the local distribution system. The North American Electric Reliability Corporation (NERC) is the federally designated U.S. Electric Reliability Organization, and is overseen by FERC. NERC has responsibility for conducting reliability assessments and developing and enforcing mandatory standards to ensure the reliability of the bulk power system—i.e., facilities and control systems necessary for operating the transmission network and certain generation facilities needed for reliability. NERC develops reliability standards collaboratively through a deliberative process involving utilities and others in the industry, which are then sent to FERC for approval. These standards include critical infrastructure protection standards for protecting electric utility-critical and cyber-critical assets. FERC has responsibility for reviewing and approving the reliability standards or directing NERC to modify them.

In addition, the Energy Independence and Security Act of 2007[6] established federal policy to support the modernization of the electricity grid and required actions by a number of federal agencies, including the National Institute of Standards and Technology (NIST), FERC, and the

---

[6]Pub. L. No. 110-140 (Dec. 19, 2007).

Department of Energy. With regard to cybersecurity, the act required NIST and FERC to take the following actions:

- NIST was to coordinate development of a framework that includes protocols and model standards for information management to achieve interoperability of smart grid devices and systems. As part of its efforts to accomplish this, NIST planned to identify cybersecurity standards for these systems and also identified the need to develop guidelines for organizations such as electric companies on how to securely implement smart grid systems. In January 2011,[7] we reported that NIST had identified 11 standards involving cybersecurity that support smart grid interoperability and had issued a first version of a cybersecurity guideline.[8]
- FERC was to adopt standards resulting from NIST's efforts that it deemed necessary to ensure smart grid functionality and interoperability. However, according to FERC officials, the statute did not provide specific additional authority to allow FERC to require utilities or manufacturers of smart grid technologies to follow these standards. As a result, any standards identified and developed through the NIST-led process are voluntary unless regulators use other authorities to indirectly compel utilities and manufacturers to follow them.

The Electricity Grid Is Potentially Vulnerable to an Evolving Array of Cyber-Based Threats

Threats to systems supporting critical infrastructure—which includes the electricity industry and its transmission and distribution systems—are evolving and growing. In February 2011, the Director of National Intelligence testified that, in the past year, there had been a dramatic increase in malicious cyber activity targeting U.S. computers and networks, including a more than tripling of the volume of malicious software since 2009.[9] Different types of cyber threats from numerous

---

[7]GAO-11-117.

[8]NIST Special Publication 1108, *NIST Framework and Roadmap for Smart Grid Interoperability Standards*, Release 1.0, January 2010 and NIST Interagency Report 7628, *Guidelines for Smart Grid Cyber Security*, August 2010.

[9]Director of National Intelligence, *Statement for the Record on the Worldwide Threat Assessment of the U.S. Intelligence Community*, statement before the Senate Select Committee on Intelligence (Feb. 16, 2011).

sources may adversely affect computers, software, networks, organizations, entire industries, or the Internet. Cyber threats can be unintentional or intentional. Unintentional threats can be caused by software upgrades or maintenance procedures that inadvertently disrupt systems. Intentional threats include both targeted and untargeted attacks from a variety of sources, including criminal groups, hackers, disgruntled employees, foreign nations engaged in espionage and information warfare, and terrorists. Table 1 shows common sources of cyber threats.

Table 1: Sources of Cybersecurity Threats

Threat source	Description
Bot-network operators	Bot-net operators use a network, or bot-net, of compromised, remotely controlled systems to coordinate attacks and to distribute phishing schemes, spam, and malware attacks. The services of these networks are sometimes made available on underground markets (e.g., purchasing a denial-of-service attack or services to relay spam or phishing attacks).
Criminal groups	Criminal groups seek to attack systems for monetary gain. Specifically, organized criminal groups use spam, phishing, and spyware/malware to commit identity theft, online fraud, and computer extortion. International corporate spies and criminal organizations also pose a threat to the United States through their ability to conduct industrial espionage and large-scale monetary theft and to hire or develop hacker talent.
Hackers	Hackers break into networks for the thrill of the challenge, bragging rights in the hacker community, revenge, stalking, monetary gain, and political activism, among other reasons. While gaining unauthorized access once required a fair amount of skill or computer knowledge, hackers can now download attack scripts and protocols from the Internet and launch them against victim sites. Thus, while attack tools have become more sophisticated, they have also become easier to use. According to the Central Intelligence Agency, the large majority of hackers do not have the requisite expertise to threaten difficult targets such as critical U.S. networks. Nevertheless, the worldwide population of hackers poses a relatively high threat of an isolated or brief disruption causing serious damage.
Insiders	The disgruntled organization insider is a principal source of computer crime. Insiders may not need a great deal of knowledge about computer intrusions because their knowledge of a target system often allows them to gain unrestricted access to cause damage to the system or to steal system data. The insider threat includes contractors hired by the organization, as well as careless or poorly trained employees who may inadvertently introduce malware into systems.
Nations	Nations use cyber tools as part of their information-gathering and espionage activities. In addition, several nations are aggressively working to develop information warfare doctrine, programs, and capabilities. Such capabilities enable a single entity to have a significant and serious impact by disrupting the supply, communications, and economic infrastructures that support military power—impacts that could affect the daily lives of citizens across the country. In his January 2012 testimony, the Director of National Intelligence stated that, among state actors, China and Russia are of particular concern.
Phishers	Individuals or small groups execute phishing schemes in an attempt to steal identities or information for monetary gain. Phishers may also use spam and spyware or malware to accomplish their objectives.
Spammers	Individuals or organizations distribute unsolicited e-mail with hidden or false information in order to sell products, conduct phishing schemes, distribute spyware or malware, or attack organizations (e.g., a denial of service).

Threat source	Description
Spyware or malware authors	Individuals or organizations with malicious intent carry out attacks against users by producing and distributing spyware and malware. Several destructive computer viruses and worms have harmed files and hard drives, including the Melissa Macro Virus, the Explore.Zip worm, the CIH (Chernobyl) Virus, Nimda, Code Red, Slammer, and Blaster.
Terrorists	Terrorists seek to destroy, incapacitate, or exploit critical infrastructures in order to threaten national security, cause mass casualties, weaken the economy, and damage public morale and confidence. Terrorists may use phishing schemes or spyware/malware in order to generate funds or gather sensitive information.

Source: GAO analysis based on data from the Director of National Intelligence, Department of Justice, Central Intelligence Agency, and the Software Engineering Institute's CERT® Coordination Center.

These sources of cyber threats make use of various techniques, or exploits that may adversely affect computers, software, a network, an organization's operation, an industry, or the Internet itself. Table 2 shows common types of cyber exploits.

Table 2: Types of Cyber Exploits

Type of exploit	Description
Cross-site scripting	An attack that uses third-party web resources to run script within the victim's web browser or scriptable application. This occurs when a browser visits a malicious website or clicks a malicious link. The most dangerous consequences occur when this method is used to exploit additional vulnerabilities that may permit an attacker to steal cookies (data exchanged between a web server and a browser), log key strokes, capture screen shots, discover and collect network information, and remotely access and control the victim's machine.
Denial-of-service	An attack that prevents or impairs the authorized use of networks, systems, or applications by exhausting resources.
Distributed denial-of-service	A variant of the denial-of-service attack that uses numerous hosts to perform the attack.
Logic bombs	A piece of programming code intentionally inserted into a software system that will cause a malicious function to occur when one or more specified conditions are met.
Phishing	A digital form of social engineering that uses authentic-looking, but fake, e-mails to request information from users or direct them to a fake website that requests information.
Passive wiretapping	The monitoring or recording of data, such as passwords transmitted in clear text, while they are being transmitted over a communications link. This is done without altering or affecting the data.
Structured Query Language (SQL) injection	An attack that involves the alteration of a database search in a web-based application, which can be used to obtain unauthorized access to sensitive information in a database.
Trojan horse	A computer program that appears to have a useful function, but also has a hidden and potentially malicious function that evades security mechanisms by, for example, masquerading as a useful program that a user would likely execute.
Virus	A computer program that can copy itself and infect a computer without the permission or knowledge of the user. A virus might corrupt or delete data on a computer, use e-mail programs to spread itself to other computers, or even erase everything on a hard disk. Unlike a computer worm, a virus requires human involvement (usually unwitting) to propagate.

Type of exploit	Description
War driving	The method of driving through cities and neighborhoods with a wireless-equipped computer– sometimes with a powerful antenna–searching for unsecured wireless networks.
Worm	A self-replicating, self-propagating, self-contained program that uses network mechanisms to spread itself. Unlike computer viruses, worms do not require human involvement to propagate.
Zero-day exploit	An exploit that takes advantage of a security vulnerability previously unknown to the general public. In many cases, the exploit code is written by the same person who discovered the vulnerability. By writing an exploit for the previously unknown vulnerability, the attacker creates a potent threat since the compressed timeframe between public discoveries of both makes it difficult to defend against.

Source: GAO analysis of data from the National Institute of Standards and Technology, United States Computer Emergency Readiness Team, and industry reports.

Electricity Grid Faces Cybersecurity Vulnerabilities

The potential impact of these threats is amplified by the connectivity between information systems, the Internet, and other infrastructures, creating opportunities for attackers to disrupt critical services, including electrical power. In addition, the increased reliance on IT systems and networks also exposes the electric grid to potential and known cybersecurity vulnerabilities. These vulnerabilities include

- an increased number of entry points and paths that can be exploited by potential adversaries and other unauthorized users;
- the introduction of new, unknown vulnerabilities due to an increased use of new system and network technologies;
- wider access to systems and networks due to increased connectivity; and
- an increased amount of customer information being collected and transmitted, providing incentives for adversaries to attack these systems and potentially putting private information at risk of unauthorized disclosure and use.

In May 2008, we reported that the corporate network of the Tennessee Valley Authority—the nation's largest public power company, which generates and distributes power in an area of about 80,000 square miles in the southeastern United States—contained security weaknesses that could lead to the disruption of control systems networks and devices connected to that network.[10] We made 19 recommendations to improve the implementation of information security program activities for the

---

[10]GAO, *Information Security: TVA Needs to Address Weaknesses in Control Systems and Networks*, GAO-08-526 (Washington, D.C.: May 21, 2008).

control systems governing the Tennessee Valley Authority's critical infrastructures and 73 recommendations to address specific weaknesses in security controls. The Tennessee Valley Authority concurred with the recommendations and has taken steps to implement them.

We and others have also reported that smart grid and related systems have known cyber vulnerabilities. For example, cybersecurity experts have demonstrated that certain smart meters can be successfully attacked, possibly resulting in disruption to the electricity grid. In addition, we have reported that control systems used in industrial settings such as electricity generation have vulnerabilities that could result in serious damages and disruption if exploited.[11] Further, in 2007, the Department of Homeland Security, in cooperation with the Department of Energy, ran a test that demonstrated that a vulnerability commonly referred to as "Aurora" had the potential to allow unauthorized users to remotely control, misuse, and cause damage to a small commercial electric generator. Moreover, in 2008, the Central Intelligence Agency reported that malicious activities against IT systems and networks have caused disruption of electric power capabilities in multiple regions overseas, including a case that resulted in a multicity power outage.[12] As government, private sector, and personal activities continue to move to networked operations, the threat will continue to grow.

Reported Incidents Illustrate the Potential Impact of Cyber Threats

Cyber incidents continue to affect the electricity industry. For example, the Department of Homeland Security's Industrial Control Systems Cyber Emergency Response Team recently noted that the number of reported cyber incidents affecting control systems of companies in the electricity sector increased from 3 in 2009 to 25 in 2011. In addition, we and others have reported[13] that cyber incidents can affect the operations of energy facilities, as the following examples illustrate:

- **Smart meter attacks**. In April 2012, it was reported that sometime in 2009 an electric utility asked the FBI to help it investigate widespread incidents of power thefts through its smart meter deployment. The report indicated that the miscreants hacked into the smart meters to

---

[11]GAO-07-1036.

[12]The White House, *Cyberspace Policy Review: Assuring a Trusted and Resilient Information and Communications Infrastructure* (Washington, D.C.: May 29, 2009).

[13]GAO-07-1036 and GAO-12-92.

change the power consumption recording settings using software available on the Internet.
- **Phishing attacks directed at energy sector**. The Department of Homeland Security's Industrial Control Systems Cyber Emergency Response Team reported that, in 2011, it deployed incident response teams to an electric bulk provider and an electric utility that had been victims of broader phishing attacks. The team found three malware samples and detected evidence of a sophisticated threat actor.
- **Stuxnet.** In July 2010, a sophisticated computer attack known as Stuxnet was discovered. It targeted control systems used to operate industrial processes in the energy, nuclear, and other critical sectors. It is designed to exploit a combination of vulnerabilities to gain access to its target and modify code to change the process.
- **Browns Ferry power plant.** In August 2006, two circulation pumps at Unit 3 of the Browns Ferry, Alabama, nuclear power plant failed, forcing the unit to be shut down manually. The failure of the pumps was traced to excessive traffic on the control system network, possibly caused by the failure of another control system device.
- **Northeast power blackout.** In August 2003, failure of the alarm processor in the control system of FirstEnergy, an Ohio-based electric utility, prevented control room operators from having adequate situational awareness of critical operational changes to the electrical grid. When several key transmission lines in northern Ohio tripped due to contact with trees, they initiated a cascading failure of 508 generating units at 265 power plants across eight states and a Canadian province.
- **Davis-Besse power plant**. The Nuclear Regulatory Commission confirmed that in January 2003, the Microsoft SQL Server worm known as Slammer infected a private computer network at the idled Davis-Besse nuclear power plant in Oak Harbor, Ohio, disabling a safety monitoring system for nearly 5 hours. In addition, the plant's process computer failed, and it took about 6 hours for it to become available again.

Actions Have Been Taken to Secure the Electricity Grid, but Challenges Remain

Multiple entities have taken steps to help secure the electricity grid, including NERC, NIST, FERC, and the Departments of Homeland Security and Energy. NERC has performed several activities that are intended to secure the grid. It has developed eight critical infrastructure standards for protecting electric utility-critical and cyber-critical assets.

The standards established requirements for the following key cybersecurity-related controls: critical cyber asset identification, security management controls, personnel and training, electronic "security perimeters," physical security of critical cyber assets, systems security management, incident reporting and response planning, and recovery plans for critical cyber assets. In December 2011, we reported that NERC's eight cyber security standards, along with supplementary documents, were substantially similar to NIST guidance applicable to federal agencies.[14]

NERC also has published security guidelines for companies to consider for protecting electric infrastructure systems, although such guidelines are voluntary and typically not checked for compliance. For example, NERC's June 2010 *Security Guideline for the Electricity Sector: Identifying Critical Cyber Assets* is intended to assist entities in identifying and developing a list of critical cyber assets as described in the mandatory standards. NERC also has enforced compliance with mandatory cybersecurity standards through its Compliance Monitoring and Enforcement Program, subject to FERC review. NERC has assessed monetary penalties for violations of its cyber security standards.

NIST, in implementing its responsibilities under the Energy Independence and Security Act of 2007 with regard to standards to achieve interoperability of smart grid systems, planned to identify cybersecurity standards for these systems. In January 2011, we reported[15] that it had identified 11 standards involving cybersecurity that support smart grid interoperability and had issued a first version of a cybersecurity guideline.[16] NIST's cybersecurity guidelines largely addressed key cybersecurity elements, such as assessment of cybersecurity risks and identification of security requirements (i.e., controls); however, its guidelines did not address an important element essential to securing smart grid systems—the risk of attacks using both cyber and physical means.[17] NIST officials said that they intended to update the guidelines to address this and other missing elements they identified, but their plan and

---

[14] GAO-12-92.

[15] GAO-11-117.

[16] NIST Special Publication 1108, *NIST Framework and Roadmap for Smart Grid Interoperability Standards*, Release 1.0, January 2010 and NIST Interagency Report 7628, *Guidelines for Smart Grid Cyber Security*, August 2010.

[17] GAO-11-117.

schedule for doing so were still in draft form. We recommended that NIST finalize its plan and schedule for incorporating missing elements, and NIST officials agreed. We are currently working with officials to determine the status of their efforts to address these recommendations.

FERC also has taken several actions to help secure the electricity grid. For example, it reviewed and approved NERC's eight critical infrastructure protection standards in 2008. Since then, in its role of overseeing the development of reliability standards, the commission has directed NERC to make numerous changes to standards to improve cybersecurity protections. However, according to the FERC Chairman's February 2012 letter in response to our report on electricity grid modernization, many of the outstanding directives have not been incorporated into the latest versions of the standards. The Chairman added that the commission would continue to work with NERC to incorporate the directives. In addition, FERC has authorized NERC to enforce mandatory reliability standards for the bulk power system, while retaining its authority to enforce the same standards and assess penalties for violations. We reported in January 2011 that FERC also had begun reviewing initial smart grid standards identified as part of NIST efforts. However, in July 2011, the commission declined to adopt the initial smart grid standards identified as a part of the NIST efforts, finding that there was insufficient consensus to do so.

The Department of Homeland Security has been designated by federal policy as the principal federal agency to lead, integrate, and coordinate the implementation of efforts to protect cyber-critical infrastructures and key resources. Under this role, the Department's National Cyber Security Division's Control Systems Security Program has issued recommended practices to reduce risks to industrial control systems within and across all critical infrastructure and key resources sectors, including the electricity subsector. For example, in April 2011, the program issued the *Catalog of Control Systems Security: Recommendations for Standards Developers*, which is intended to provide a detailed listing of recommended controls from several standards related to control systems.[18] The program also manages and operates the Industrial Control Systems Cyber Emergency Response Team to respond to and analyze control-systems-related incidents, provide onsite support for incident response and forensic analysis, provide situational awareness in the form of actionable

---

[18]DHS, National Cyber Security Division, Control Systems Security Program, *Catalog of Control Systems Security: Recommendations for Standards Developers* (April 2011).

intelligence, and share and coordinate vulnerability information and threat analysis through information products and alerts. For example, it reported providing on-site assistance to six companies in the electricity subsector, including a bulk electric power provider and multiple electric utilities, during 2009-2011.

The Department of Energy is the lead federal agency which is responsible for coordinating critical infrastructure protection efforts with the public and private stakeholders in the energy sector, including the electricity subsector. In this regard, we have reported that officials from the Department's Office of Electricity Delivery and Energy Reliability stated that the department was involved in efforts to assist the electricity sector in the development, assessment, and sharing of cybersecurity standards.[19] For example, the department was working with NIST to enable state power producers to use current cybersecurity guidance. In May 2012, the department released the *Electricity Subsector Cybersecurity Risk Management Process*.[20] The guideline is intended to ensure that cybersecurity risks for the electric grid are addressed at the organization, mission or business process, and information system levels. We have not evaluated this guide.

Challenges to Securing Electricity Systems and Networks

In our January 2011 report, we identified a number of key challenges that industry and government stakeholders faced in ensuring the cybersecurity of the systems and networks that support our nation's electricity grid.[21] These included the following:

- *There was a lack of a coordinated approach to monitor whether industry follows voluntary standards.* As mentioned above, under the Energy Independence and Security Act of 2007, FERC is responsible for adopting cybersecurity and other standards that it deems necessary to ensure smart grid functionality and interoperability. However, FERC had not developed an approach coordinated with other regulators to monitor, at a high level, the extent to which industry will follow the voluntary smart grid standards it adopts. There had been initial efforts by regulators to share views, through, for

---

[19]GAO-12-92.

[20]U.S. Department of Energy, *Electricity Subsector Cybersecurity Risk Management Process*, DOE/OE-0003 (Washington, D.C.: May 2012).

[21]GAO-11-117.

example, a collaborative dialogue between FERC and the National Association of Regulatory Utility Commissioners, which had discussed the standards-setting process in general terms. Nevertheless, according to officials from FERC and the National Association of Regulatory Utility Commissioners, FERC and the state public utility commissions had not established a joint approach for monitoring how widely voluntary smart grid standards are followed in the electricity industry or developed strategies for addressing any gaps. Moreover, FERC had not coordinated in such a way with groups representing public power or cooperative utilities, which are not routinely subject to FERC's or the states' regulatory jurisdiction for rate setting. We noted that without a good understanding of whether utilities and manufacturers are following smart grid standards, it would be difficult for FERC and other regulators to know whether a voluntary approach to standards setting is effective or if changes are needed.[22]

- *Aspects of the current regulatory environment made it difficult to ensure the cybersecurity of smart grid systems.* In particular, jurisdictional issues and the difficulties associated with responding to continually evolving cyber threats were a key regulatory challenge to ensuring the cybersecurity of smart grid systems as they are deployed. Regarding jurisdiction, experts we spoke with expressed concern that there was a lack of clarity about the division of responsibility between federal and state regulators, particularly regarding cybersecurity. While jurisdictional responsibility has historically been determined by whether a technology is located on the transmission or distribution system, experts raised concerns that smart grid technology may blur these lines. For example, devices such as smart meters deployed on parts of the grid traditionally subject to state jurisdiction could, in the aggregate, have an impact on those parts of the grid that federal regulators are responsible for— namely the reliability of the transmission system.

---

[22]In an order issued on July 19, 2011, FERC reported that it had found insufficient consensus to institute a rulemaking proceeding to adopt smart grid interoperability standards identified by NIST as ready for consideration by regulatory authorities. While FERC dismissed the rulemaking, it encouraged utilities, smart grid product manufacturers, regulators, and other smart grid stakeholders to actively participate in the NIST interoperability framework process to work on the development of interoperability standards and to refer to that process for guidance on smart grid standards. Despite this result, we believe our recommendations to FERC in GAO-11-117, with which FERC concurred, remain valid and should be acted upon as consensus is reached and standards adopted.

There was also concern about the ability of regulatory bodies to respond to evolving cybersecurity threats. For example, one expert questioned the ability of government agencies to adapt to rapidly evolving threats, while another highlighted the need for regulations to be capable of responding to the evolving cybersecurity issues. In addition, our experts expressed concern with agencies developing regulations in the future that are overly specific in their requirements, such as those specifying the use of a particular product or technology. Consequently, unless steps are taken to mitigate these challenges, regulations may not be fully effective in protecting smart grid technology from cybersecurity threats.

- *Utilities were focusing on regulatory compliance instead of comprehensive security.* The existing federal and state regulatory environment creates a culture within the utility industry of focusing on compliance with cybersecurity requirements, instead of a culture focused on achieving comprehensive and effective cybersecurity. Specifically, experts told us that utilities focus on achieving minimum regulatory requirements rather than designing a comprehensive approach to system security. In addition, one expert stated that security requirements are inherently incomplete, and having a culture that views the security problem as being solved once those requirements are met will leave an organization vulnerable to cyber attack. Consequently, without a comprehensive approach to security, utilities leave themselves open to unnecessary risk.

- *There was a lack of security features built into smart grid systems.* Security features are not consistently built into smart grid devices. For example, experts told us that certain currently available smart meters had not been designed with a strong security architecture and lacked important security features, including event logging[23] and forensics capabilities that are needed to detect and analyze attacks. In addition, our experts stated that smart grid home area networks—used for managing the electricity usage of appliances and other devices in the home—did not have adequate security built in, thus increasing their vulnerability to attack. Without securely designed smart grid systems, utilities may lack the capability to detect and analyze attacks, increasing the risk that attacks will succeed and utilities will be unable to prevent them from recurring.

---

[23]Event logging is a capability of an IT system to record events occurring within an organization's systems and networks, including those related to computer security.

- *The electricity industry did not have an effective mechanism for sharing information on cybersecurity and other issues.* The electricity industry lacked an effective mechanism to disclose information about cybersecurity vulnerabilities, incidents, threats, lessons learned, and best practices in the industry. For example, our experts stated that while the electricity industry has an information sharing center, it did not fully address these information needs. In addition, President Obama's May 2009 cyberspace policy review also identified challenges related to cybersecurity information sharing within the electric and other critical infrastructure sectors and issued recommendations to address them.[24] According to our experts, information regarding incidents such as both unsuccessful and successful attacks must be able to be shared in a safe and secure way to avoid publicly revealing the reported organization and penalizing entities actively engaged in corrective action. Such information sharing across the industry could provide important information regarding the level of attempted cyber attacks and their methods, which could help grid operators better defend against them. If the industry pursued this end, it could draw upon the practices and approaches of other industries when designing an industry-led approach to cybersecurity information sharing. Without quality processes for information sharing, utilities will not have the information needed to adequately protect their assets against attackers.

- *The electricity industry did not have metrics for evaluating cybersecurity.* The electricity industry was also challenged by a lack of cybersecurity metrics, making it difficult to measure the extent to which investments in cybersecurity improve the security of smart grid systems. Experts noted that while such metrics[25] are difficult to develop, they could help compare the effectiveness of competing solutions and determine what mix of solutions combine to make the most secure system. Furthermore, our experts said that having metrics would help utilities develop a business case for cybersecurity by helping to show the return on a particular investment. Until such metrics are developed, there is increased risk that utilities will not invest in security in a cost-effective manner, or have the information

---

[24]The White House, *Cyberspace Policy Review: Assuring a Trusted and Resilient Information and Communications Infrastructure* (Washington, D.C.: May 29, 2009).

[25]Metrics can be used for, among other things, measuring the effectiveness of cybersecurity controls for detecting and blocking cyber attacks.

needed to make informed decisions on their cybersecurity investments.

To address these challenges, we made recommendations in our January 2011 report. To improve coordination among regulators and help Congress better assess the effectiveness of the voluntary smart grid standards process, we recommended that the Chairman of FERC develop an approach to coordinate with state regulators and with groups that represent utilities subject to less FERC and state regulation to (1) periodically evaluate the extent to which utilities and manufacturers are following voluntary interoperability and cybersecurity standards and (2) develop strategies for addressing any gaps in compliance with standards that are identified as a result of this evaluation. We also recommended that FERC, working with NERC as appropriate, assess whether commission efforts should address any of the cybersecurity challenges identified in our report. FERC agreed with these recommendations.

Although FERC agreed with these recommendations, they have not yet been implemented. According to the FERC Chairman, given the continuing evolution of standards and the lack of sufficient consensus for regulatory adoption, commission staff believe that coordinated monitoring of compliance with standards would be premature at this time, and that this may change as new standards are developed and deployed in industry. We believe that it is still important for FERC to improve coordination among regulators and that consensus is reached on standards. We will continue to monitor the status of its efforts to address these recommendations.

In summary, the evolving and growing threat from cyber-based attacks highlights the importance of securing the electricity industry's systems and networks. A successful attack could result in widespread power outages, significant monetary costs, damage to property, and loss of life. The roles of NERC and FERC remain critical in approving and disseminating cybersecurity guidance and enforcing standards, as appropriate. Moreover, more needs to be done to meet challenges facing the industry in enhancing security, particularly as the generation, transmission, and distribution of electricity comes to rely more on emerging and sophisticated technology.

Chairman Bingaman, Ranking Member Murkowski, and Members of the Committee, this concludes my statement. I would be happy to answer any questions you may have at this time.

Contact and Acknowledgments

If you have any questions regarding this statement, please contact Gregory C. Wilshusen at (202) 512-6244 or wilshuseng@gao.gov or David C. Trimble, Director, Natural Resources and Environment Team, at (202) 512-3841 or trimbled@gao.gov. Other key contributors to this statement include Michael Gilmore, Anjalique Lawrence, and Jon R. Ludwigson (Assistant Directors), Paige Gilbreath, Barbarol James, Lee McCracken, and Dana Pon.

Appendix I: Related GAO Products

Cybersecurity: Threats Impacting the Nation. GAO-12-666T. Washington, D.C.: April 24, 2012.

Cybersecurity: Challenges in Securing the Modernized Electricity Grid, GAO-12-507T. Washington, D.C.: February 28, 2012.

Critical Infrastructure Protection: Cybersecurity Guidance Is Available, but More Can Be Done to Promote Its Use. GAO-12-92. Washington, D.C.: December 9, 2011.

High-Risk Series: An Update. GAO-11-278. Washington, D.C.: February 2011.

Electricity Grid Modernization: Progress Being Made on Cybersecurity Guidelines, but Key Challenges Remain to Be Addressed. GAO-11-117. Washington, D.C.: January 12, 2011.

Cybersecurity: Continued Attention Needed to Protect Our Nation's Critical Infrastructure. GAO-11-865T. Washington, D.C.: July 26, 2011.

Critical Infrastructure Protection: Key Private and Public Cyber Expectations Need to Be Consistently Addressed. GAO-10-628. Washington, D.C.: July 15, 2010.

Cyberspace: United States Faces Challenges in Addressing Global Cybersecurity and Governance. GAO-10-606. Washington, D.C.: July 2, 2010.

Cybersecurity: Continued Attention Is Needed to Protect Federal Information Systems from Evolving Threats. GAO-10-834T. Washington, D.C.: June 16, 2010.

Critical Infrastructure Protection: Update to National Infrastructure Protection Plan Includes Increased Emphasis on Risk Management and Resilience. GAO-10-296. Washington, D.C.: March 5, 2010.

Cybersecurity: Progress Made but Challenges Remain in Defining and Coordinating the Comprehensive National Initiative. GAO-10-338. Washington, D.C.: March 5, 2010.

Cybersecurity: Continued Efforts Are Needed to Protect Information Systems from Evolving Threats. GAO-10-230T. Washington, D.C.: November 17, 2009.

Defense Critical Infrastructure: Actions Needed to Improve the Identification and Management of Electrical Power Risks and

Vulnerabilities to DOD Critical Assets. GAO-10-147. Washington, D.C.: October 23, 2009.

Critical Infrastructure Protection: Current Cyber Sector-Specific Planning Approach Needs Reassessment. GAO-09-969. Washington, D.C.: September 24, 2009.

National Cybersecurity Strategy: Key Improvements Are Needed to Strengthen the Nation's Posture. GAO-09-432T. Washington, D.C.: March 10, 2009.

Electricity Restructuring: FERC Could Take Additional Steps to Analyze Regional Transmission Organizations' Benefits and Performance. GAO-08-987. Washington, D.C.: September 22, 2008.

Information Security: TVA Needs to Address Weaknesses in Control Systems and Networks. GAO-08-526. Washington, D.C.: May 21, 2008.

Critical Infrastructure Protection: Multiple Efforts to Secure Control Systems Are Under Way, but Challenges Remain. GAO-07-1036. Washington, D.C.: September 10, 2007.

Cybercrime: Public and Private Entities Face Challenges in Addressing Cyber Threats. GAO-07-705. Washington, D.C.: June 22, 2007.

Meeting Energy Demand in the 21st Century: Many Challenges and Key Questions. GAO-05-414T. Washington, D.C.: March 16, 2005.

This is a work of the U.S. government and is not subject to copyright protection in the United States. The published product may be reproduced and distributed in its entirety without further permission from GAO. However, because this work may contain copyrighted images or other material, permission from the copyright holder may be necessary if you wish to reproduce this material separately.

GAO's Mission	The Government Accountability Office, the audit, evaluation, and investigative arm of Congress, exists to support Congress in meeting its constitutional responsibilities and to help improve the performance and accountability of the federal government for the American people. GAO examines the use of public funds; evaluates federal programs and policies; and provides analyses, recommendations, and other assistance to help Congress make informed oversight, policy, and funding decisions. GAO's commitment to good government is reflected in its core values of accountability, integrity, and reliability.
Obtaining Copies of GAO Reports and Testimony	The fastest and easiest way to obtain copies of GAO documents at no cost is through GAO's website (www.gao.gov). Each weekday afternoon, GAO posts on its website newly released reports, testimony, and correspondence. To have GAO e-mail you a list of newly posted products, go to www.gao.gov and select "E-mail Updates."
Order by Phone	The price of each GAO publication reflects GAO's actual cost of production and distribution and depends on the number of pages in the publication and whether the publication is printed in color or black and white. Pricing and ordering information is posted on GAO's website, http://www.gao.gov/ordering.htm. Place orders by calling (202) 512-6000, toll free (866) 801-7077, or TDD (202) 512-2537. Orders may be paid for using American Express, Discover Card, MasterCard, Visa, check, or money order. Call for additional information.
Connect with GAO	Connect with GAO on Facebook, Flickr, Twitter, and YouTube. Subscribe to our RSS Feeds or E-mail Updates. Listen to our Podcasts. Visit GAO on the web at www.gao.gov.
To Report Fraud, Waste, and Abuse in Federal Programs	Contact: Website: www.gao.gov/fraudnet/fraudnet.htm E-mail: fraudnet@gao.gov Automated answering system: (800) 424-5454 or (202) 512-7470
Congressional Relations	Katherine Siggerud, Managing Director, siggerudk@gao.gov, (202) 512-4400, U.S. Government Accountability Office, 441 G Street NW, Room 7125, Washington, DC 20548
Public Affairs	Chuck Young, Managing Director, youngc1@gao.gov, (202) 512-4800 U.S. Government Accountability Office, 441 G Street NW, Room 7149 Washington, DC 20548

Please Print on Recycled Paper.

www.ingramcontent.com/pod-product-compliance
Lightning Source LLC
Chambersburg PA
CBHW081421170526

45166CB00010B/3429